1. Introduction

In the remote community of Dhalinybuy, we have been invited to meet some of the elders, perhaps to hear some ancient Aboriginal stories of the sky. As we approach, we see a group of men and children sitting on a mat surrounding an elderly man. His chest bears the proud scars of a fully initiated Yolngu man, and his bearing and charisma tell you that he is a leader of his people.

Figure 1: Mathulu Munyarryun

After we are invited to sit with them, the old man is introduced as "Mathulu." We are astonished. Mathulu Munyarryun? Famous ceremonial leader, custodian of ancient stories of the sky?

In our quest over the last few years to understand the astronomy of the Yolngu people, we often hear tales of Venus as the Morning Star. But when we ask people about the corresponding Evening Star, we are told to ask Mathulu, who is the custodian of the "Evening Star story". At last, in this remote Yolngu community of Dhalinybuy in North-East Arnhem Land*, we meet him. But more of Mathulu later.

There used to be about 400 different Aboriginal cultures in Australia. Each had its own language, stories, and beliefs, although most were centred on the idea that the world was created in the "Dreaming" by ances-

* A map can be found on page 26.

tral spirits, whose presence can still be seen both on the land and in the sky. These spirits taught humans how to live, and have left a user guide to life in their songs and stories. Many of these stories are reflected in the patterns in the sky. Aboriginal people searching for meaning in the dark skies of ancient Australia would notice that particular stars were visible only at certain times of the year, and so the night sky would naturally be an important chapter of this user guide.

The 50,000 year-old Aboriginal cultures are believed to be the oldest continuous cultures in the world. Since the night sky seems to play an important role in them, it is sometimes said that they include the earliest traces of astronomy. But astronomy is more than just stories – it means a quest to understand the patterns in the sky, and the motion and eclipses of the Sun, Moon, and planets. So the goal of our research project is to understand the importance of Astronomy in Aboriginal Cultures, and ultimately perhaps answer the question "Were the Australian Aboriginal people the world's first astronomers?"

Tragically, many of the Aboriginal cultures have been severely damaged, or even wiped out, since the arrival of Europeans. So we have divided our project into two parts. One focuses on people like those in Arnhem Land, who have managed to keep their culture pretty well intact, and still conduct initiation ceremonies where knowledge is passed on from one generation to the next. The other focus is on South-Eastern Australia, where Aboriginal cultures were almost wiped out. But their legacy of beautiful rock engravings and stone arrangements provides us with clues to their cultures.

Of course, we also depend on the research of those who have gone before us. Especially moving is the detailed record of Wardaman astronomy[1] written by Bill Yidumduma Harney and Hugh Cairns. Other important works include those by Ros Haynes[2], Dianne Johnson[3], Philip Clarke[4], and other significant studies discussed below.

Much traditional Aboriginal knowledge is sacred, and can only be given to those who have been properly initiated. Naturally we respect this,

Emu Dreaming

An Introduction to Australian Aboriginal Astronomy

by

Ray & Cilla Norris

This booklet is dedicated to the hundreds of thousands of Indigenous Australians who lost their lives as a result of the British occupation of Australia in 1788.

www.EmuDreaming.com

ISBN 978-0-9806570-0-5

First published July 2009 by Emu Dreaming, Sydney, Australia

Emu Dreaming

An Introduction to Australian Aboriginal Astronomy

First published July 2009 by
Emu Dreaming
PO Box 4335
North Rocks
NSW 2151
Australia
www.EmuDreaming.com

Reprinted April 2010

ISBN 978-0-9806570-0-5

Thanks to Barnaby Norris for permission to use his "Emu in the Sky" image on the covers and in the Figures. High-quality prints of this image are available from www.emuinthesky.com

and are careful not to intrude where we are not welcome. But Aboriginal elders have encouraged us to learn a little of the knowledge, and pass it on so that others may understand something of the richness and complexity of their culture. This booklet is part of the result.

2. The Milky Way and the Emu

On an Australian winter evening, the Milky Way dominates the sky – a broad band of light stretching high across the heavens. Within this glowing band lie dark patches and channels. These are clouds of dust in which new stars are being born.

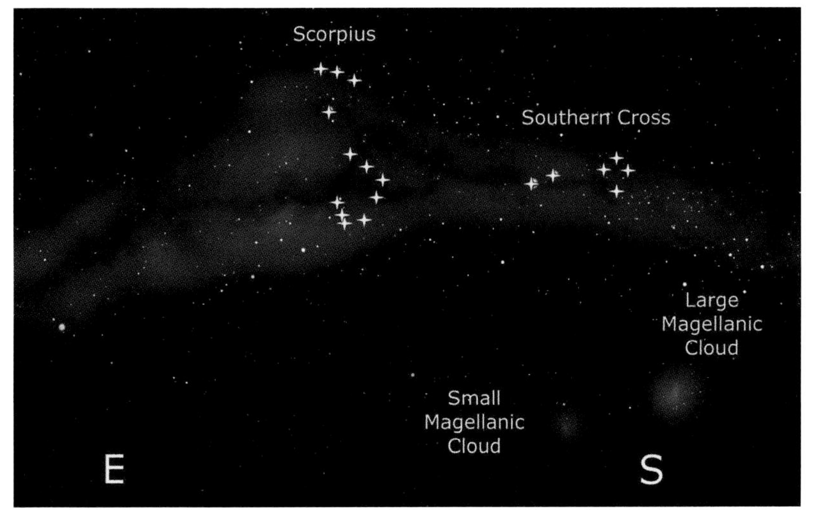

Figure 2: The night sky as seen from Sydney at about 9pm on June 30, looking South-East.

Our European ancestors explained the patterns of stars in terms of constellations of Greek gods. The Aboriginal people living in Australia thousands of years ago also interpreted these patterns, but with a difference. Whereas the Europeans saw only the patterns of stars, the Aboriginal people also saw meaning in the dark patches.

On an evening between May and August, drive out to a remote spot in the bush, away from streetlights, and look up. Directly above you stretches the spectacular band of the Milky Way reaching across the sky from horizon to horizon. Astronomers tell us that we are looking into the disc of our Galaxy, and that the stream of light comes from millions of distant stars. Yolngu people from the Top End of Australia tell us that it is a mighty river, and either side of it can be seen the campfires of their ancestors. Desert people tell us that it is the Rainbow Serpent.

Figure 3: Crossing a creek on the Central Arnhem Highway. The clear inviting waters of the creek, fringed by Pandanus, are deceptive. Like most other creeks along the 700km dirt road, the water is home to crocodiles with an appetite for humans. Not a good place for the engine to stall.

High up to the South you will see the Southern Cross. The central desert people say this is the footprint of an Eagle[5]. Look at the two bright stars (the "pointers") to the lower left of the cross. The leftmost one is called Alpha Centauri by astronomers. Unlike most of the stars you see, which are hundreds of light years* away, Alpha Centauri is a mere four light years away and is the closest star to the Earth (apart from the Sun, of course). The two pointers are the Eagle's throwing sticks. Now find the small dark patch of sky between the pointers and the Southern Cross. That dark cloud is called the Coalsack by astronomers, who say the darkness is caused by dense clouds of interstellar dust clumped together. Deep within the clouds lies a stellar nursery in which new stars are born. Central desert people say that the Coalsack is the Eagle's nest[6].

The Boorong people of Victoria have a different story about this part of the sky. They say that the Southern Cross is a tree, and the star at the top of the cross is a possum, who has just been chased up the tree by an emu[7], represented by the Coalsack. The emu in turn is being chased by two brothers who are the pointers[8].

Figure 4: A tracing of the emu engraving in Ku-ring-gai Chase National Park is a close fit to the Emu in the sky.

To many Aboriginal groups, the Coalsack is part of the best known Aboriginal constellation – the Emu in the Sky. The Coalsack is the head of the emu. Stretching away to its left you should be able to see its long dark neck, round body (near Scorpius),

* Astronomers measure distances in light years. One light year is the distance travelled by a beam of light in a year, which is about 10 trillion km.

and finally (towards the horizon) the legs. This Emu in the Sky features in the songs and stories of Aboriginal groups right across Australia, from Western Australia to New South Wales, although it's not universal, and detailed interpretations differ. It's a spectacular sight – far better than the contrived European constellations that most of us grew up with. Once you've seen it on a dark night in the Australian bush, the Milky Way will never look the same again.

Just North of Sydney lies Ku-ring-gai Chase National Park, where the Guringai people lived until the British arrived in 1788. The Guringai people were known for their beautiful rock engravings, showing people, animals, creator spirits, and strange symbols whose meaning is lost. Shortly after the First Fleet arrived in Sydney, Governor Phillip explored the region and noted the friendliness of the Guringai people. Within two years, most had died of smallpox. A few decades later, white settlers and their diseases had driven out the remaining Guringai people, but nobody thought to ask them about the engravings. Now, only the engravings themselves can tell their story.

Amongst these sacred works of art, close to the Elvina Track, is a finely engraved emu. Its legs trail behind it, in a position that would be unnatural for a real emu, but, as Sydney academic Hugh Cairns pointed out a few years ago, is exactly that of the Emu in the Sky. This might sound like wild speculation until you notice that, astonishingly, the time when the Emu in the Sky stands above her portrait, in the correct orientation, is just the time when real-life emus are laying their eggs. It seems very likely that this engraving is a picture of the Emu in the Sky rather than a real emu.

Next to the emu portrait is an engraving of a strange half-man with a club foot and impressive head-dress, who is believed to be Daramulan, a creator-hero from the Dreaming of the Guringai people. He is related to Baiame, a creator-hero found in many Aboriginal cultures across the South-East of Australia. Only fragments of information have survived about the culture of the Guringai people, but perhaps we can learn a little about them from their rock art.

Figure 5: The emu in the sky above her engraving in Ku-Ring-Gai Chase National Park, at the time of the year when emus are laying their eggs. Photograph by Barnaby Norris. Prints of this photo, which won a New Scientist "Eureka" prize, are available on www.emuinthesky.com

3. The Magellanic Clouds

Figure 6: The Large Magellanic Cloud.

Drive back out to your favourite remote spot in the bush, away from street-lights, and look high in the Southern Sky. Above you are thousands of stars, each typically a few hundred light years away.

These "local" stars together make up our Galaxy, which we call the Milky Way. Almost everything you can see in the sky with the naked eye is within the Milky Way. However, you can also see two fuzzy clouds, which were discovered in 1520 by the explorer Magellan, and so are called the Magellanic Clouds*. The Magellanic Clouds are much, much further away (about 160,000 light years), and are, together with the Andromeda Nebula in the far North, the only things outside our Galaxy that you can see with the naked eye. Those fuzzy glowing clouds, each consisting of billions of stars, are Galaxies like our own Milky Way, but smaller. They are circling our Galaxy and one day may merge with it.

Aboriginal groups in Arnhem Land say that the Magellanic Clouds are the campfires of an old couple, who live some distance away from the river of the Milky Way along which the younger people camp. The couple are too weak to gather food for themselves, so other star people hunt fish and gather food from the Milky Way for them to cook on

---

* In the Southern States of Australia, the two Magellanic Clouds are always above the horizon, often high in the sky, although they are sometimes low, even below the tree-line. They are less easy to see in the North of Australia, as they are lower in the sky, and dip below the horizon at some times of the year.

their fire[9]. Aboriginal groups in Victoria say that the Magellanic Clouds are a pair of Brolgas[10,11], while others say that they are two brothers who come to Earth to collect the spirits of dead people [12,13].

4. Sun, Moon, and Eclipses

In most Aboriginal cultures, the Moon is male and the Sun is female. For example, the Yolngu people tell how Walu, the Sun-woman, lights a fire each morning, bringing us dawn[14]. As she decorates herself with red ochre, some spills onto the clouds, creating a red sunrise. Picking up a stringy-bark tree, she lights it from the fire, and then travels across the sky from east to west carrying her blazing torch, creating daylight. On reaching the western horizon, again some of the ochre dusts the clouds to give the red sunset. She extinguishes her torch, and starts the long journey underground back to the morning camp in the east.

The Yolngu people tell how Ngalindi, the Moon-man, was originally a fat and lazy man (the full Moon), who demanded that his wives and sons feed him. When his sons refused to do so, he beat and killed them. When his wives found out, they attacked him with their axes, chopping bits out of him. As a result, he became thinner (the waning Moon) and

Figure 7. The phases of the Moon.

tried to escape by following the Sun. But it was in vain, and he died of his injuries. After remaining dead for 3 days (the new Moon), he rose again, growing fat and round (the waxing Moon), until, after two weeks his wives attacked him again. To this day, the cycle continues every month.

When Ngalindi first died, it was normal for people and animals to return to life after dying, but he laid a curse so that, henceforth, only he could return to life. For everyone else, death would be final. So Ngalindi is responsible for bringing death to the world. A similar story is found in Aboriginal groups across Australia, although the details differ, and the cause of the Moon-man's death varies from group to group, and even from clan to clan within a group. Other versions known by initiated men detail the taboos that the Moon-man broke and which caused his death, and which also contain powerful symbolic knowledge.

The Yolngu stories also explain the association between the Moon and tides. Sailors know that high tides follow the Moon, and that the highest tides occur at Full Moon and New Moon. Astronomers explain this in terms of the Moon's gravity. The traditional Yolngu explanation is that, when the tides are high, water fills the Moon as it rises through the horizon, causing a full Moon. Later, when the Moon is only half-full, the tides become lower. A week later, the water runs out of the Moon, raising the tides, leaving the Moon empty for three days. On the next rising tide, the Moon refills. So this emptying and filling also results in the waxing and waning of the Moon. Such stories show that, although the mechanics are a little different from our modern version, traditional Aboriginal cultures contain a detailed knowledge of the motion of the Moon, and its effects on the Earth.

Figure 8: Ngalindi, as a very fat man.

These stories are also a good illustration of two different explanations of the same phenomenon – the waxing and waning of the Moon being caused either by Ngalindi being attacked by his wives, or by the Moon filling with water. There is no inconsistency here – it is common for the stories to overlap. Sometimes they are from different language groups, sometimes from different clans, but often several overlapping stories occur in the same clan.

Astronomers tell us that solar eclipses are caused when the Moon gets between the Sun and the Earth, blocking the Sun's light from those of us on Earth. A lunar eclipse, on the other hand, happens when the Earth gets between the Sun and the Moon, casting the Earth's shadow

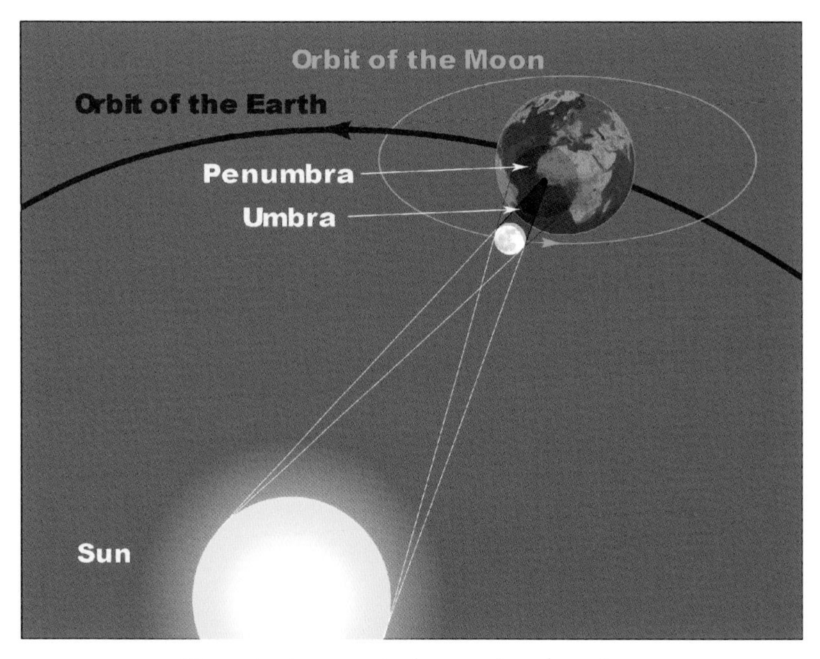

Figure 9: Astronomical view of a solar eclipse.

on the Moon, making the Moon go dark. Total lunar eclipses are quite common, and are visible every few years. Solar eclipses, however, are rare, and a total solar eclipse occurs in any one location only once every 370 years, so most people will never see one in their lifetime. You might therefore expect that solar eclipses, when the Sun goes out in the middle of the day, would be startling, and perhaps terrifying, to a culture without written records from previous generations.

So it's surprising that Aboriginal stories contain explanations of both solar and lunar eclipses. The people of North-West Arnhem Land say a solar eclipse happens when the Sun-woman is hidden by the Moon-man as he makes love to her[15]. On the other hand, a lunar eclipse occurs when the Moon-man is pursued and overtaken by the Sun-woman. These two stories demonstrate that traditional Aboriginal people had already figured out that the Sun and Moon move on different paths across the sky, and that eclipses occur on those rare occasions when the two bodies meet at the intersection of their paths.

This realisation is found in several other language groups. The eccentric Irishwoman Daisy Bates, living in the desert in her starched blouse and lace-up boots, recounted[16] primly how, during the solar eclipse of 1922, the Wirangu people told her that the eclipse was caused when the Sun and Moon became "guri-arra – husband and wife together."

Modern astronomers understand how the motion of the Moon is far more complex than that of the Sun, moving to the North and South of the Sun during the course of the month, and this knowledge is reflected in Aboriginal stories of the Sun-woman making a steady path across the sky while the Moon-man zigzags around trying to escape her attentions. When she catches him, then of course they make love, causing an eclipse.

The stories of solar eclipses imply an amazing continuity of culture and learning over many generations. Without written records, the knowledge of solar eclipses must have been passed from generation to generation by oral tradition, carefully memorised and passed to the next generation by individuals who would never see an eclipse in their lifetime.

Earlier, we discussed the "Emu in the Sky" rock engraving of the Guringai people. These same people left behind thousands of beautiful sacred rock engravings depicting the Dreaming ancestors, images of the animals and fish that abound in and around the Park, and some strange symbols that we don't understand. A recurring motif in the Ku-ring-gai engravings is a crescent, which has been interpreted by archaeologists as a boomerang. But a closer look suggests a different explanation. Figure 10 shows a man and woman reaching up to a boomerang in the sky. But is it a boomerang? Why would a giant boomerang be sailing over their heads? And boomerangs don't usually have pointed ends, and generally have two straight lengths rather than a single curved crescent. It seems much more likely that the shape represents the crescent Moon.

But if these shapes are Moons, then why is the Moon shown with the two horns pointing down, since that configuration is seen only in the afternoon or morning when the Sun is already high in the sky, and the Moon barely visible?

One answer is that it might depict an eclipse, when the Moon (or the Sun) can be seen as a crescent with its horns pointing downwards (I am indebted to Chris Douglass for this suggestion). In Figure 10, the man and woman are overlapping, so one is partly obscured by the other*. Such carefully-drawn obscurations are unusual in these rock carvings, and in this case may well represent an eclipse.

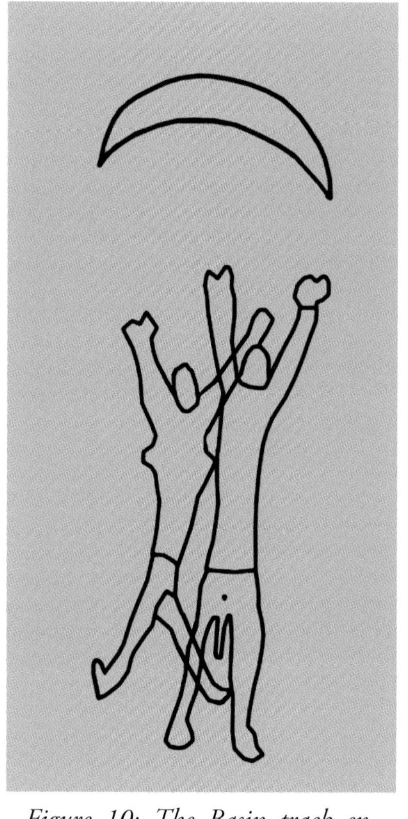

Figure 10: The Basin track engraving of a man and woman and crescent.

Near Woy-Woy on the central coast of New South Wales is a rock engraving of a creator-spirit named Bulgandry, who may be related to Baiame. In one hand is a disc, and in the other is a crescent shape. Nowadays, the crescent is badly eroded and barely visible, but fortunately we have a drawing made by W. D. Campbell[17] in the 1890's which shows it clearly.

---

* Earlier sketches of this engraving show the man obscuring the woman, but in the present-day engraving, the lines of each figure continue through the other, so it isn't clear which figure is in front.

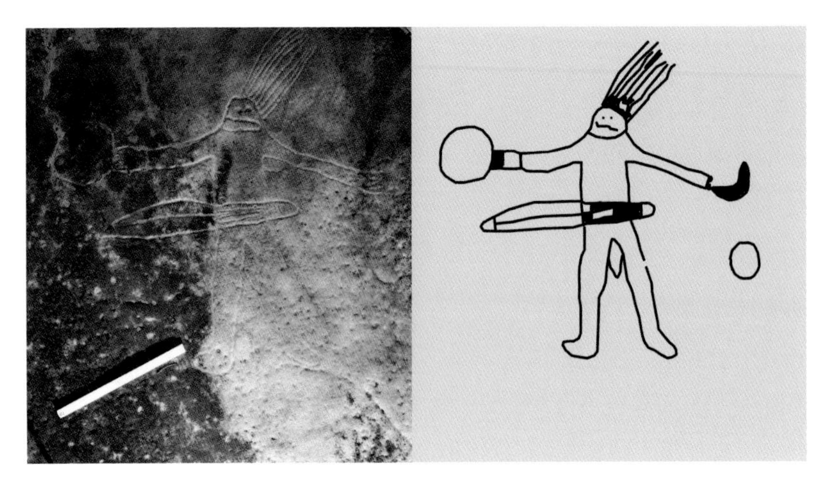

Figure 11: The engraving of Bulgandry, near Woy Woy, NSW

Archaeologists label these objects as a shield and a boomerang. However, the shield is the wrong shape and has none of the markings that are usually shown on shield engravings in this area, and the crescent is Moon-shaped rather than boomerang-shaped. It is tempting to speculate that this engraving may refer to a Dreaming story about the creation of the Sun and Moon. But we must acknowledge that this is no more than speculation. To make sense of these engravings will take years of detective work, amassing clues and piecing them together. But we now know that astronomy is an important part of the culture of Aboriginal groups elsewhere in Australia. So astronomy may turn out to be an important component of the Sydney rock engravings.

5. Calendars and Constellations

Aboriginal calendars tend to be more complex than European calendars, and those in the north of Australia are often based on six seasons. These seasons are distinguished both by the appearance of certain plants on the ground, and by the appearance of certain stars or constellations in the sky. For example, the Pitjantjatjara people of the Central Desert say that the rising of the Pleiades (Seven Sisters) in the dawn sky in May heralds the start of winter[18]. Perhaps even more importantly, the time when a star or constellation first becomes visible each year (the heliacal rising) tells people when it's time to move camp, to a place where a different food source is coming in to season. For example, when the Mallee-fowl constellation (Lyra) appears in March, the Boorong people of Victoria know that the Mallee-fowl are about to build their nests. When she disappears in October, they know the eggs are laid and are ready to be collected[19]. Similarly, the appearance of Scorpius once signalled to the Yolngu people of Arnhem Land the imminent arrival of the Macassan (Indonesian) fisherman, who brought goods to be traded[20].

Figure 12: The constellation of Orion, also often known as "the saucepan" in Australia. Towards the top of the photo is a fuzzy blob known as the Orion Nebula, where new stars are being born.

Some of the most familiar European constellations also have Aboriginal names. For example, the familiar constellation of Orion (also known in Australia as the saucepan), visible in the summer evenings, gets its European name from the hunter named Orion in Greek mythology. The only problem with this Greek interpretation is that the constellation of Orion appears upside down in Australia, so the hunter Orion is standing on his head. There has to be a better Australian interpretation!

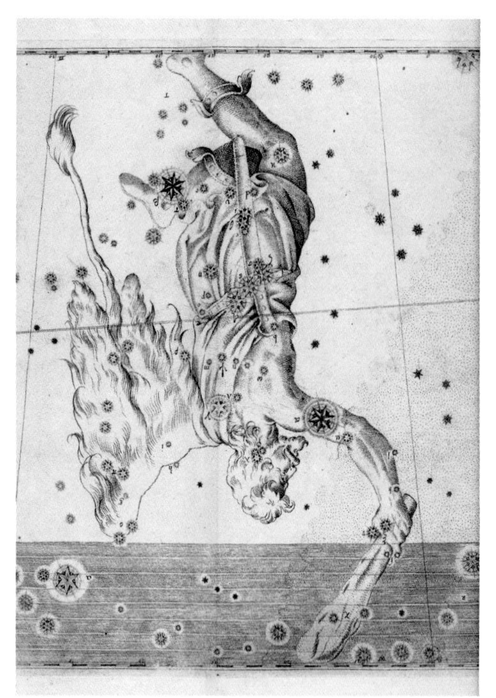

Figure 13: The constellation of Orion, as interpreted as a Greek hunter in Johann Bayer's Uranometria, but inverted as seen from Australia.

And of course there is. Aboriginal people thousands of years ago also gave a name to this constellation. Strangely, like the Greek story, it tends to represent a hunter, or group of hunters, all across Australia. For example, in Yolngu language the constellation is called Djulpan. The story[21] goes that three brothers of the kingfish clan went out hunting and fishing, but failed to catch anything, except for kingfish. Because they were from the kingfish clan, Yolngu law forbade them from eating kingfish – it would be like cannibalism. But one brother eventually got so hungry that he started eating a kingfish. The Sun, Walu, saw this and became so angry that she blew the brothers and their canoe up into the sky, where you can still see them. The Orion nebula is the fish, still trailing behind them in the water on its line.

In Greek mythology, the Pleiades (or Seven Sisters) were a group of sisters who were chased by Orion. Oddly, almost exactly the same story is found in Aboriginal groups across Australia: the Pleiades are seen as six or seven sisters who are chased by the hunter (or hunters) in Orion.

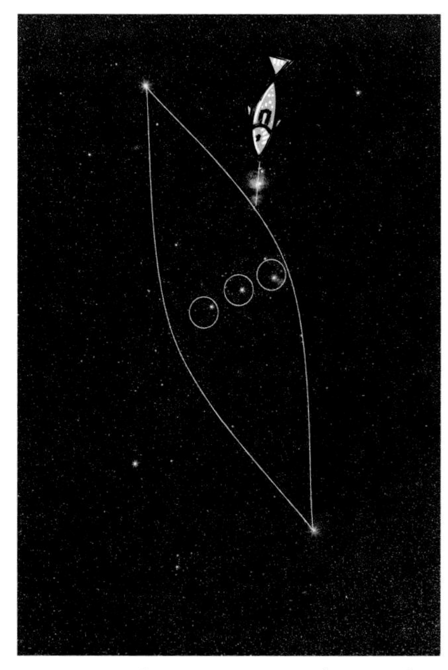

Figure 14: Orion, or Djulpan. The three stars in the centre of the canoe are the three brothers, and the king-fish still trails in the water behind them

Although some anthropologists once concluded from this similarity that there must have been early contact between Aboriginal and European people, this is now thought unlikely. Instead, a sort of cultural convergent evolution seems to have taken place, in which Aboriginal people independently devised a similar story to that of the Europeans. Perhaps this isn't too surprising. Orion, with its massive pattern of dominant bright stars, has to be the most macho constellation in the sky – a veritable gorilla marching across the summer sky. The Pleiades, by contrast, are a pretty little group of starlets that twinkle endearingly. Add to these characteristics the fact that, as the sky rotates, Orion appears to follow the Pleiades into the West, and all the elements of the story are there.

An interesting alternative explanation is that stories like these popping up in completely unrelated cultures may have ancient roots. Perhaps these stories originate in our common ancestors who moved out of Africa some hundred thousand years ago,

Fig 15: The Pleiades, or seven sisters,

eventually to produce all modern humans including both Aboriginal Australians and Europeans.

6. Morning Star, Evening Star

The silvery glow of Venus, the Morning Star, must be one of the most spectacular sights in the sky. It is extremely important to Yolngu people, who call her Banumbirr, and tell how in the Dreaming she came across the ocean from the East, from the Island of the Dead, Baralku. As she crossed the shoreline near Yirrkala, Banumbirr named and created the animals and places. She continued travelling westwards across the land, naming waterholes, rivers, and mountains. The path that she followed is now traced by a "songline" which is still commemorated in Yolngu songs and ceremonies, and provides a navigational route across the Top End of Australia[22].

Unlike the planets of Mars and Jupiter, which travel right across the sky, the planet Venus is only ever seen close to the Sun, either as a Morning Star rising just before dawn, or as an Evening Star just after sunset. Modern astronomers explain this in terms of the orbits of the planets around the Sun. Yolngu people, probably thousand of years ago, also observed this phenomenon, and came up with a different explanation.

When Venus rises before dawn, a rope is said to hang below her, which connects her to the island of Baralku. This rope is said to prevent her moving too far from the Sun.

Yolngu people say that if you look carefully, you can see this rope. We didn't understand this, until it was pointed out to us. Yolngu people seem to be referring to the faint glowing line in the sky that astronomers call the zodiacal light, caused by extraterrestrial dust in the solar system. Although difficult to see in most parts of the world, it is easily visible in the clear dark skies and low latitude of northern Australia.

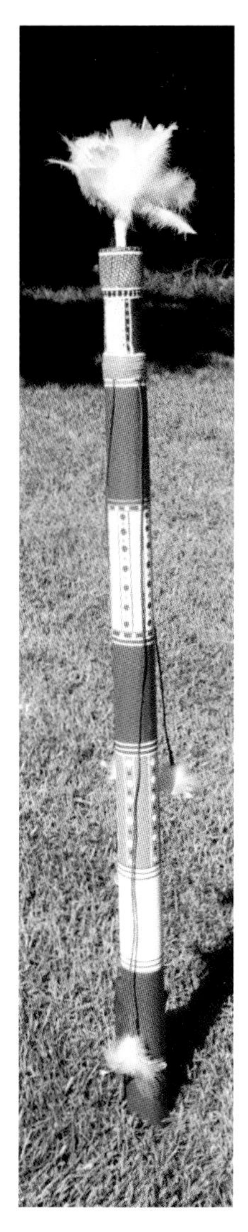

Some Yolngu clans still hold an important "Morning Star Ceremony" as part of the funeral process. This ceremony starts at dusk and continues to a climax as Banumbirr (Venus) rises before dawn. During the ceremony, a "Morning Star Pole" is used to help the participants communicate with their ancestors, with Banumbirr's help. The messages are said to be carried along the rope which connects Banumbirr to Baralku, where the ancestors live.

We can learn two important things from the Morning Star ceremony. One is that Yolngu tradition includes the knowledge that Venus never moves far from the Sun, which is explained by a rope binding Venus to the position of the rising Sun. The other is that since the Morning-Star ceremony needs to be planned, and Venus rises before dawn only at certain times (roughly every 1.5 years), Yolngu people also keep track of the path of Venus well enough to predict when to hold the Morning Star Ceremony.

What of the Evening Star? For half the time Venus rises before the Sun, and is then called the Morning Star. For the other half, Venus rises after the Sun, and is not visible in the glare of the morning daylight. However, at those times of the year it also sets after the Sun, and then is visible as a bright star in the West after sunset – the Evening Star. So do Yolngu people have a corresponding story about the Evening Star?

Fig. 16: A morning star pole. The tuft of Magpie-goose feathers at the top represents Venus, and the other feathers represent nearby stars, and other clans.

They do, but the story[23] is puzzling. It says that when the Evening Star, named Djurrpun, is visible in the evening, that's the time to harvest Raika, the Lotus bulbs that grow in Arnhem Land. But that doesn't seem to make sense - Venus wouldn't be any good for telling you when to harvest Raika, as its setting time changes from year to year, reappearing in the same apparent position only after a period of 584 days. So this is why we were desperately seeking Mathulu at the start of this book. Mathulu, we were told, was the owner of this story, and the only person who could explain it.

After trying, unsuccessfully for a couple of years, to meet him, we were invited to tag along with some teachers on their weekly visit to the remote community of Dhalinybuy, 200km away. A lunar eclipse was due to occur that evening, so it would be a good opportunity to teach the kids about astronomy in general, and eclipses in particular. When we arrived in Dhalinybuy, we helped the teachers get their stuff ready, and also asked one of the Yolngu teachers, Gurumin, if it would be possible to speak to the elders.

Late in the afternoon, Gurumin told us that the older men were ready to speak to us about their stories. We followed him across the community, to a group of men sitting on a mat. Following Yolngu custom, we waited nearby until invited over. That's when we found that, at last, we were speaking to Mathulu Munyarryun! Our hearts leapt – perhaps at last we would understand the mystery of the Evening Star.

Mathulu was more comfortable speaking in Yolngu, so a younger man, Banul, translated for us. The story started: "A lady went out to a waterhole, and she sat collecting raika nuts…"

We'd read this story before, about the Evening Star, Djurrpun, but it was wonderful to hear it from Mathulu in this exotic setting. He continued. "When Djurrpun sets just after the Sun, we know that raika, the nuts from the rushes in the river, are ready to be harvested."

So now we were getting to the nub of the story. If this story was accurate, then Djurrpun couldn't be Venus, as its setting time changes from year to year. We asked them which star was Djurrpun, and Banul promised to show us after dark.

Figure 17: Banul and the Evening Star Rope

And then the conversation took an unexpected twist. From a bag, Banul produced a long rope.

"This is a Laka – an Evening Star Rope", Mathulu told us.

We were amazed. We'd spent the last couple of years reading the literature on Yolngu culture, and had never before heard of the Laka. This was probably new knowledge as far as Western academia was concerned.

"It's a line of stars in the sky", explained Mathulu, "and when the first star sets just after sunset, that's the time for the women to collect the raika nuts."

He let us examine the rope, made of pandanus twine, twisted together with possum fur and orange lorikeet feathers. Woven in to the rope were the yellow-white marbles of the raika nuts.

"This Laka is a memorial to my grandmother," continued Mathulu, "and we used it at her funeral to send her spirit off to the evening star. Like this."

Together, he and Banul demonstrated how a line of mourners held the rope on their heads, joining their spirits to that of their grandmother as they say farewell. Walking back, we asked Banul which was the Evening Star.

"That one," he pronounced confidently, pointing at the star Spica. Yes that fits. Spica sets behind the Sun in October, just before the Raika harvest. So the anthropology books which said Djurrpun was Venus were plain wrong. Another puzzle solved.

7. Astronomical Measurements

Perhaps the most difficult challenge is to test what we call the "Stonehenge Hypothesis": is there any evidence that traditional Aboriginal people made careful observations, kept records, or built structures to point to the rising and setting places of heavenly bodies?

In South Australia, on the dreamy banks of the Murray River, north of Adelaide, the Nganguraku people engraved images of the Sun and Moon at a site called "Ngaut Ngaut". Next to these images are a series of dots and lines carved in the rock, which, say the traditional owners, show the "cycles of the Moon". But such knowledge is usually passed through generations from father to son, and from elder to novice at initiation ceremonies. Since Christian missionaries tried to stifle the Nganguraku language and culture over a hundred years ago, only this fragment of information has survived, and it is not known exactly what the symbols mean. Any more precise information about these engravings has evaporated, and the engravings themselves have so far resisted attempts to decode them. Per-

Figure 18: The carvings at Ngaut Ngaut, said to represent lunar cycles.

haps one day we'll succeed, but for now we must label it as intriguing, but uncertain, evidence of Aboriginal measurements of astronomical bodies.

What about structures that mark rising and setting positions? The Wurdi Youang stone arrangement in Victoria is an impressive egg-shaped ring of stones, about 50 metres across, with its major axis almost exactly East-West. The ring is dominated by an eye-catching group of three waist-high stones at its highest point, the western apex. The Wathaurung people built it long before European settlement, which destroyed their culture along with the knowledge of how it was used.

Figure 19: The view across the Wurdi Youang stone arrangement showing the positions of the setting Sun at the solstices and equinoxes.

Attention was drawn to this site by John Morieson who pointed out[24] that if you look between the two largest stones, small outlying stones mark the position on the horizon where the Sun sets on midwinter's day, on midsummer's day, and at the equinox.

These orientations have recently been confirmed by our new survey[25], but some doubts remain. First, the alignments are accurate to no more than a few degrees, raising the possibility that they may have occurred by chance. Second, although the stones of the circle are large and difficult to move, the outliers are small and may have been disturbed.

On the other hand, our new survey has discovered a new piece of evidence which supports Morieson's suggestion. Not only do the outliers point towards the setting places of the midsummer and midwinter Sun, but two straight sections of the ring point in these same directions. So this one site has two independent pieces of evidence to support the idea

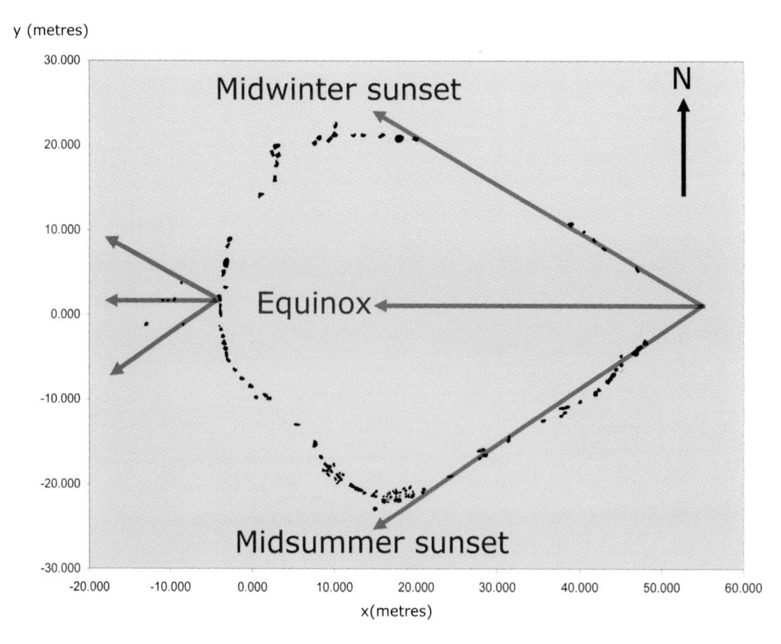

Figure 20: Plan of the Wurdi Youang stone arrangement showing the directions to the setting Sun at the solstices and equinoxes.

that it was intended to point towards the place on the horizon where the Sun sets on these important days.

To summarise the findings at Wurdi Youang, we have some strong evidence that the site was designed to show the position of sunset on special days, thus perhaps marking out a calendar. But the evidence is not yet strong enough to be indisputable, and the case would be much stronger if we could find a second site aligned to the same directions.

A search for such a site has so far been unsuccessful – we know of other stone arrangements in Victoria, but most have been badly damaged or destroyed. On the other hand, we can still see that some point towards the cardinal points (north, south, east, and west) and so it is clear that the Aboriginal people in this area were familiar with, and interested in, these compass directions. Work is continuing to see if we can find more evidence for astronomical alignments in these stone arrangements.

8. Conclusion

What can we conclude from the evidence presented so far? Certainly that traditional Aboriginal people were interested in the sky, and used the stars for time-keeping and navigation. They knew that eclipses happened when the paths of the Sun and Moon intersected, and they knew how the Moon was related to the tides. They had a deep and extensive knowledge of the sky, and of the motion of the celestial bodies across it. Perhaps they were even making accurate measurements of the rising and setting places of the Sun.

But we are only in the early stages of this study, and it is likely that far more lies undiscovered. I suspect that this is only the very beginning of a long and fascinating journey.

More details can be found in our published research papers[26] and on http://www.EmuDreaming.com

Appendix: Motivation for this Work

Why should we be interested in Aboriginal Astronomy? Certainly, one factor is personal curiosity. But it goes much deeper. First, if we succeed in uncovering ancient Aboriginal astronomies, perhaps we can give back to the Aboriginal people some of that culture which our European ancestors destroyed when they occupied Australia two hundred years ago.

Second, there is still a gulf of misunderstanding between Indigenous and non-Indigenous Australians. Over a beer at a barbecue, our mate, Dave, sniggers some insensitive and plain stupid comments about "secret women's business", not understanding that this is something real, and sacred, and important to many of his fellow-Australians. Dave isn't bad, but he doesn't begin to understand the cultures of Aboriginal Australians, and the complex issues now facing them.

How can we help Dave to understand Indigenous Australian culture? Many of the esoteric Indigenous stories and tales are just too different for Dave to understand. On the other hand, Dave loves the bark paintings that he sees in the tourist shops, and the Didgeridoo music he hears on Circular Quay. All these art forms have successfully forged a bridge of understanding across our cultures.

We hope that, like music and art, astronomy can build an important bridge of understanding between Indigenous and non-Indigenous Australians. We all share the same sky, and can't help being awed by the beauty and mystery of the glorious Milky Way stretching across the unknowable black heavens above us. We all love swapping stories about it. In doing so, this project aims to promote a greater appreciation of the depth and richness of Indigenous Australian cultures.

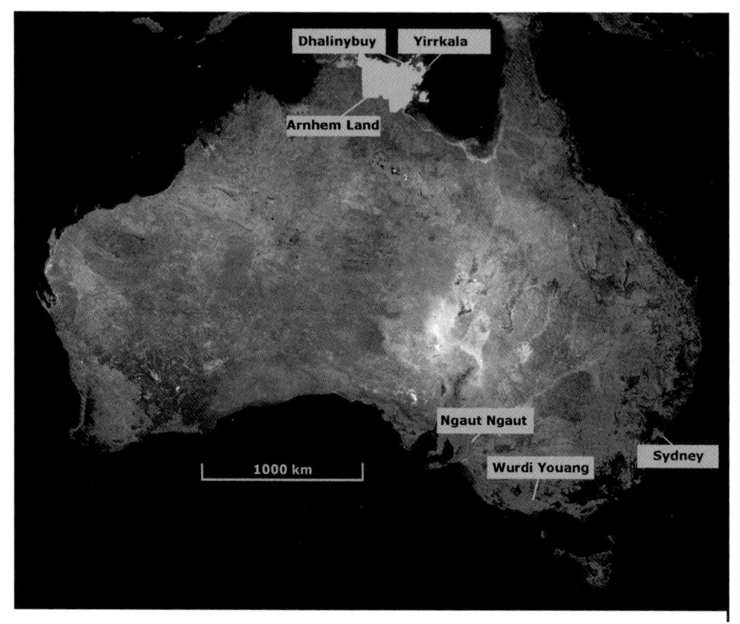

Figure 21: Map of Australia showing the places mentioned in the text.

The Authors

Ray Norris is an astrophysicist at the CSIRO Australia Telescope National Facility. Born in England, he obtained an MA in theoretical physics at Cambridge, followed by a PhD and postdoc in radio-astronomy at Manchester, while also studying the astronomy of ancient standing stones. In 1983, Ray and his family moved to Australia where he joined CSIRO, and now researches the formation of the first galaxies in the Universe, and also the astronomy of Aboriginal Australians. He was appointed as an Adjunct Professor at the Macquarie University Dept. of Indigenous Studies (Warawara) in 2008.

Cilla Norris has been an artist, high school teacher, veterinary nurse, wildlife sanctuary guide, and wild-life carer. As well as working with Ray on Aboriginal Astronomy, she is known as an authority on the care and rehabilitation of possums, and writes and teaches about possums to the NSW Wildlife Information and Rescue Service (WIRES) and other groups.

Ray and Cilla Norris

Acknowledgements

This booklet is dedicated to the hundreds of thousands of Indigenous Australians who lost their lives as a result of the British occupation of Australia in 1788.

We acknowledge and pay our respects to Aboriginal elders, past and present. We would like to thank the Yolngu elders and communities of Arnhem Land, the Buku-Larrnggay Mulka Centre, and Yirrkala Dhanbul Community Association, and of course our friends in Yirrkala for their help, hospitality, and for their permission to tell their stories. We especially thank Djapirri Mununggirritj for her friendship, guidance, and patience with our stupid Napaki questions!

We are also grateful to the other Indigenous groups who have welcomed us onto their land, and acknowledge the Bidjigal, Darkinjung, Darug, Eora, and Guringai people, on whose land this book was written, and whose stories and art initially motivated much of this work. We also thank Sydney Metropolitan Land Council, the NSW National Parks and Wildlife Service, and the traditional custodians of Ngaut Ngaut and Wurdi Youang, for continuing to preserve and care for these important legacies of human history and achievement. We thank the organisations who have sponsored parts of this work either financially or in kind: CSIRO, Gove Amateur Astronomers, Macquarie University, Rio Tinto Alcan, and the Yothu Yindi Foundation.

We are indebted to our colleagues Hugh Cairns, John Clegg, Paul Curnow, Kristina Everett, Serena Fredrick, Duane Hamacher, Ray Johnston, John Morieson, Adele Pring, Clive Ruggles, and especially Ian MacLean, for their inspiration, help, and collaboration. Our warm thanks go to Bruce and Janette Mcnaughton for their hospitality and friendship in Nhulunbuy. We also thank Barnaby and Tamasin Norris for help with some of the surveys and photography used in this book, and give a special thanks to Barnaby for his magnificent "Emu in the Sky" Image.

Image Credits

All images are © Ray and Cilla Norris, except where stated below.

Covers and Figs. 4 & 5 (Emu in the Sky). © Barnaby Norris, reproduced with permission. See www.emuinthesky.com

Fig 2: Based on an image from Stellarium, used under the Gnu General Public License.

Fig 6: Courtesy of NASA C-141 KAO Imagery

Fig 7: Courtesy of the Royal Geographical Society, from Wikipedia under the Creative Commons License.

Fig 9: Courtesy of Sagredo, from Wikimedia Commons under the Creative Commons License.

Fig 11: Composite image made from photo taken by Ray Norris, and a sketch made by Campbell (1899).

Fig 12: Courtesy of Mouser, from Wikimedia Commons under the Creative Commons License.

Fig 13: Courtesy of the United States Naval Observatory Library.

Fig 15: Courtesy of NASA/ESA/AURA/Caltech.

Fig 16: Pole created by Richard Garrawurra. Photo by Ray Norris.

Fig 19: Composite from originals by Ray Norris and John Morieson.

Fig 21: Original satellite image courtesy of NASA.

References

[1] Cairns, H.C, & Yidumduma Harney, Bill, 2004, "Dark Sparklers", publ. Hugh Cairns, Merimbula.

[2] Haynes, R.F., Haynes, R.D., Malin, D., McGee, R.X., 1996, "Explorers of the Southern Sky", CUP, Cambridge.

[3] Johnson, Dianne, 1998, "Night Skies of Aboriginal Australia: A Noctuary", Oceania Publications, Sydney.

[4] Clarke, P.A., 2003, "Where the Ancestors Walked. Australia as an Aboriginal Landscape", Allen & Unwin, Sydney.

[5] Mountford, C.P., 1976, "Nomads of the Australian Desert", Rigby, Adelaide.

[6] Mountford, C.P., 1976, *ibid.*

[7] Stanbridge, W.E., 1857, "On the astronomy and mythology of the Aborigines of Victoria", Proc of the Philosophical Institute of Victoria, Transactions 2, 137-140.

[8] Stanbridge, W.E., 1857, *ibid.*

[9] Mountford, C. P. (1956). Records of the American-Australian Scientific Expedition to Arnhem Land. Volume 1: Art, Myth and Symbolism. Melbourne University Press.

[10] Stanbridge, W.E., 1857, *ibid.*

[11] Massola, Aldo, 1968, "Bunjil's cave", Lansdowne Press, Melbourne.

[12] Mountford, C.P., 1976, *ibid.*

[13] Bates, Daisy, 1904, "Papers of Daisy Bates", National Library of Australia (reference courtesy of Serena Fredrick).

[14] Wells, A.E.,1973, "Stars in the Sky", Rigby, Adelaide.

[15] Johnson, Dianne, 1998. *ibid.*

[16] Bates, Daisy, 1944, "The Passing of the Aborigines", John Murray, London.

[17] Campbell, W. D. 1899, "Aboriginal Carvings of Port Jackson and Broken Bay", Memoirs of the Geological Survey of NSW, Ethnographical Series No. 1, Sydney.

[18] Clarke, P.A., 2003, *ibid.*

[19] Stanbridge, W.E., 1857, *ibid.*

[20] Mountford, C.P., 1976, *ibid.*

[21] Wells, A.E.,1973, *ibid.*

[22] Personal communication from Yolngu members of the Yirrkala community.

[23] Wells, A.E.,1973, *ibid.*

[24] Morieson, J., 2003,"Solar-based Lithic Design in Victoria, Australia", in World Archaeological Congress, Washington DC, 2003.

[25] Norris, R.P., Norris, P.M., & Hamacher, D.W., 2009, in preparation.

[26] Norris, R.P., & Hamacher, D.W., 2009, "The Astronomy of Aboriginal Australia", in "The Role of Astronomy in Society and Culture", Proc. IAU Symp. No. 260, also on http://arxiv.org/abs/0906.0155v1.